This book belongs to:

_____

_____

Table of Contents

ROUNDING TO THE NEAREST THOUSAND (4 DIGIT)

Rounding a four-digit number to the nearest thousand means that you want to approximate the number to the nearest multiple of 1,000. Here's how you can do it step by step:

Identify the digit in the thousands place. This is the leftmost digit in a four-digit number.

Look at the digit immediately to the right of the thousands place digit. This is the hundreds place digit.

If the hundreds place digit is 0, 1, 2, 3, or 4, you round down the thousands place digit to the nearest thousand. If the hundreds place digit is 5, 6, 7, 8, or 9, you round up the thousands place digit to the nearest thousand.

Replace all the digits to the right of the thousands place with zeros.

Let's go through some examples:

Example 1: Round 4,723 to the nearest thousand.

The thousands place digit is 4.

The hundreds place digit is 7.

Since the hundreds place digit (7) is greater than or equal to 5, we round up the thousands place digit (4) to the nearest thousand. So, the result is:

4,723 rounded to the nearest thousand is 5,000.

Example 2: Round 3,286 to the nearest thousand.

The thousands place digit is 3.

The hundreds place digit is 2.

Since the hundreds place digit (2) is less than 5, we round down the thousands place digit (3) to the nearest thousand. So, the result is:

3,286 rounded to the nearest thousand is 3,000.

Example 3: Round 6,549 to the nearest thousand.

The thousands place digit is 6.

The hundreds place digit is 5.

Since the hundreds place digit (5) is greater than or equal to 5, we round up the thousands place digit (6) to the nearest thousand. So, the result is:

6,549 rounded to the nearest thousand is 7,000.

Rounding to the Nearest Thousand (4 Digit)

Round each number to the nearest thousand

Worksheet # 1

1) 3,901 _____

2) 3,274 _____

3) 2,600 _____

4) 4,781 _____

5) 6,209 _____

6) 4,439 _____

7) 9,173 _____

8) 6,020 _____

9) 7,358 _____

10) 1,463 _____

11) 6,493 _____

12) 7,220 _____

13) 8,400 _____

14) 2,986 _____

15) 7,993 _____

16) 8,454 _____

17) 7,593 _____

18) 2,989 _____

19) 1,886 _____

20) 3,350 _____

21) 5,106 _____

22) 6,786 _____

23) 3,980 _____

24) 2,832 _____

3

Rounding to the Nearest Thousand (4 Digit)

Round each number to the nearest thousand

Worksheet # 2

1) 7,186 _____

2) 8,928 _____

3) 4,573 _____

4) 5,889 _____

5) 8,624 _____

6) 7,106 _____

7) 7,718 _____

8) 6,107 _____

9) 7,458 _____

10) 9,545 _____

11) 6,078 _____

12) 5,016 _____

13) 4,353 _____

14) 7,532 _____

15) 8,352 _____

16) 5,308 _____

17) 1,948 _____

18) 1,541 _____

19) 2,630 _____

20) 9,105 _____

21) 1,610 _____

22) 9,795 _____

23) 6,436 _____

24) 3,259 _____

4

Rounding to the Nearest Thousand (4 Digit)

Round each number to the nearest thousand

Worksheet # 3

1) 6,104 _____

2) 2,108 _____

3) 4,201 _____

4) 2,428 _____

5) 2,207 _____

6) 9,377 _____

7) 8,269 _____

8) 4,379 _____

9) 3,220 _____

10) 2,169 _____

11) 5,758 _____

12) 1,444 _____

13) 5,452 _____

14) 5,371 _____

15) 1,303 _____

16) 2,107 _____

17) 2,180 _____

18) 5,604 _____

19) 8,495 _____

20) 8,128 _____

21) 4,558 _____

22) 5,458 _____

23) 1,954 _____

24) 9,838 _____

Rounding to the Nearest Thousand (4 Digit)

Round each number to the nearest thousand

Worksheet # 4

1) 4,218 _____

2) 6,227 _____

3) 2,459 _____

4) 1,565 _____

5) 1,458 _____

6) 5,496 _____

7) 8,686 _____

8) 8,622 _____

9) 2,183 _____

10) 1,311 _____

11) 3,582 _____

12) 6,076 _____

13) 5,454 _____

14) 7,893 _____

15) 7,306 _____

16) 3,210 _____

17) 9,400 _____

18) 9,189 _____

19) 3,968 _____

20) 7,377 _____

21) 5,211 _____

22) 3,996 _____

23) 8,780 _____

24) 4,439 _____

Rounding to the Nearest Thousand (4 Digit)

Round each number to the nearest thousand

Worksheet # 5

1) 5,782 _____

2) 8,778 _____

3) 6,071 _____

4) 6,057 _____

5) 3,977 _____

6) 9,403 _____

7) 8,630 _____

8) 2,570 _____

9) 7,343 _____

10) 5,703 _____

11) 7,756 _____

12) 2,195 _____

13) 7,445 _____

14) 6,369 _____

15) 2,435 _____

16) 3,516 _____

17) 6,898 _____

18) 6,654 _____

19) 2,867 _____

20) 7,989 _____

21) 3,518 _____

22) 9,885 _____

23) 1,599 _____

24) 7,407 _____

7

Rounding to the Nearest Thousand (4 Digit)

Round each number to the nearest thousand

Worksheet # 6

1) 3,860 _____

2) 5,321 _____

3) 2,089 _____

4) 1,582 _____

5) 1,853 _____

6) 5,440 _____

7) 4,327 _____

8) 9,177 _____

9) 2,091 _____

10) 4,655 _____

11) 7,556 _____

12) 2,698 _____

13) 8,265 _____

14) 1,183 _____

15) 5,254 _____

16) 6,895 _____

17) 8,473 _____

18) 5,899 _____

19) 3,436 _____

20) 9,965 _____

21) 8,444 _____

22) 3,545 _____

23) 9,145 _____

24) 6,098 _____

8

Rounding to the Nearest Thousand (4 Digit)

Round each number to the nearest thousand

Worksheet # 7

1) 8,863 _____

2) 5,296 _____

3) 9,765 _____

4) 8,373 _____

5) 7,598 _____

6) 9,617 _____

7) 8,482 _____

8) 9,611 _____

9) 9,592 _____

10) 1,983 _____

11) 9,420 _____

12) 8,666 _____

13) 1,161 _____

14) 7,233 _____

15) 2,315 _____

16) 7,135 _____

17) 2,459 _____

18) 2,194 _____

19) 8,862 _____

20) 6,510 _____

21) 1,488 _____

22) 2,205 _____

23) 1,613 _____

24) 3,363 _____

9

Rounding to the Nearest Thousand (4 Digit)

Round each number to the nearest thousand

Worksheet # 8

1) 8,335 _____

2) 7,211 _____

3) 7,533 _____

4) 4,398 _____

5) 4,992 _____

6) 2,251 _____

7) 6,472 _____

8) 2,103 _____

9) 7,801 _____

10) 5,356 _____

11) 3,295 _____

12) 2,213 _____

13) 9,442 _____

14) 3,943 _____

15) 9,939 _____

16) 5,744 _____

17) 5,624 _____

18) 2,078 _____

19) 3,781 _____

20) 2,977 _____

21) 4,376 _____

22) 9,508 _____

23) 6,615 _____

24) 2,852 _____

Rounding to the Nearest Thousand (4 Digit)

Round each number to the nearest thousand

Worksheet # 9

1)   9,959 _____

2)   7,111 _____

3)   4,818 _____

4)   2,006 _____

5)   1,403 _____

6)   9,609 _____

7)   3,526 _____

8)   2,168 _____

9)   5,418 _____

10) 3,493 _____

11) 5,385 _____

12) 2,557 _____

13) 9,305 _____

14) 6,632 _____

15) 3,272 _____

16) 6,281 _____

17) 1,038 _____

18) 9,919 _____

19) 9,349 _____

20) 1,508 _____

21) 9,685 _____

22) 5,387 _____

23) 6,049 _____

24) 1,216 _____

Rounding to the Nearest Thousand (4 Digit)

Round each number to the nearest thousand

Worksheet # 10

1) 4,188 _____

2) 1,200 _____

3) 7,257 _____

4) 8,758 _____

5) 1,638 _____

6) 4,241 _____

7) 6,651 _____

8) 4,046 _____

9) 8,437 _____

10) 3,394 _____

11) 7,931 _____

12) 9,795 _____

13) 5,825 _____

14) 1,124 _____

15) 1,417 _____

16) 2,414 _____

17) 9,150 _____

18) 2,900 _____

19) 3,910 _____

20) 6,505 _____

21) 8,549 _____

22) 4,914 _____

23) 6,954 _____

24) 7,159 _____

12

Rounding to the Nearest Thousand (4 Digit)

Round each number to the nearest thousand

Worksheet # 11

1) 3,547 _____

2) 1,099 _____

3) 7,544 _____

4) 5,608 _____

5) 1,229 _____

6) 7,464 _____

7) 2,151 _____

8) 7,921 _____

9) 6,346 _____

10) 1,522 _____

11) 7,769 _____

12) 3,305 _____

13) 5,160 _____

14) 5,827 _____

15) 8,149 _____

16) 4,323 _____

17) 2,247 _____

18) 9,606 _____

19) 2,854 _____

20) 1,369 _____

21) 2,976 _____

22) 5,050 _____

23) 1,206 _____

24) 3,472 _____

Rounding to the Nearest Thousand (4 Digit)

Round each number to the nearest thousand

Worksheet # 12

1) 5,804 _____

2) 9,738 _____

3) 3,707 _____

4) 7,946 _____

5) 1,796 _____

6) 3,803 _____

7) 3,467 _____

8) 5,059 _____

9) 5,039 _____

10) 4,775 _____

11) 9,859 _____

12) 9,851 _____

13) 1,852 _____

14) 4,684 _____

15) 1,263 _____

16) 2,203 _____

17) 4,280 _____

18) 1,171 _____

19) 1,539 _____

20) 8,310 _____

21) 4,648 _____

22) 3,584 _____

23) 9,461 _____

24) 1,547 _____

Rounding to the Nearest Thousand (4 Digit)

Round each number to the nearest thousand

Worksheet # 13

1) 3,669 _____

2) 5,224 _____

3) 2,917 _____

4) 1,297 _____

5) 7,316 _____

6) 2,542 _____

7) 7,199 _____

8) 5,156 _____

9) 5,729 _____

10) 5,586 _____

11) 5,514 _____

12) 8,771 _____

13) 3,306 _____

14) 5,652 _____

15) 8,678 _____

16) 2,697 _____

17) 5,499 _____

18) 5,576 _____

19) 5,631 _____

20) 8,418 _____

21) 9,127 _____

22) 5,032 _____

23) 5,313 _____

24) 5,996 _____

Rounding to the Nearest Thousand (4 Digit)

Round each number to the nearest thousand

Worksheet # 14

1) 6,714 _____

2) 2,207 _____

3) 3,024 _____

4) 5,852 _____

5) 3,739 _____

6) 6,935 _____

7) 9,917 _____

8) 2,781 _____

9) 1,439 _____

10) 6,517 _____

11) 8,825 _____

12) 6,985 _____

13) 9,176 _____

14) 3,952 _____

15) 8,633 _____

16) 7,904 _____

17) 4,120 _____

18) 5,201 _____

19) 9,784 _____

20) 3,018 _____

21) 3,447 _____

22) 8,754 _____

23) 1,498 _____

24) 8,850 _____

Rounding to the Nearest Thousand (4 Digit)

Round each number to the nearest thousand

Worksheet # 15

1) 1,958 _____

2) 5,280 _____

3) 4,770 _____

4) 4,193 _____

5) 4,656 _____

6) 6,290 _____

7) 1,643 _____

8) 3,925 _____

9) 6,212 _____

10) 6,300 _____

11) 3,445 _____

12) 2,356 _____

13) 9,588 _____

14) 5,152 _____

15) 3,214 _____

16) 9,631 _____

17) 5,080 _____

18) 4,380 _____

19) 2,642 _____

20) 4,011 _____

21) 1,524 _____

22) 8,975 _____

23) 4,445 _____

24) 9,546 _____

Rounding to the Nearest Thousand (4 Digit)

Round each number to the nearest thousand

Worksheet # 16

1) 4,342 _____

2) 8,481 _____

3) 6,698 _____

4) 3,284 _____

5) 1,622 _____

6) 9,416 _____

7) 9,042 _____

8) 9,758 _____

9) 9,760 _____

10) 3,199 _____

11) 4,334 _____

12) 1,531 _____

13) 6,234 _____

14) 1,720 _____

15) 1,844 _____

16) 6,530 _____

17) 7,292 _____

18) 2,003 _____

19) 8,579 _____

20) 7,818 _____

21) 2,342 _____

22) 7,318 _____

23) 9,553 _____

24) 7,403 _____

Rounding to the Nearest Thousand (4 Digit)

Round each number to the nearest thousand

Worksheet # 17

1) 1,166 _____

2) 4,726 _____

3) 7,204 _____

4) 4,449 _____

5) 6,007 _____

6) 5,815 _____

7) 5,044 _____

8) 4,060 _____

9) 5,866 _____

10) 1,514 _____

11) 1,357 _____

12) 5,131 _____

13) 2,933 _____

14) 7,383 _____

15) 4,015 _____

16) 1,078 _____

17) 8,313 _____

18) 5,158 _____

19) 7,732 _____

20) 9,878 _____

21) 3,598 _____

22) 5,763 _____

23) 2,031 _____

24) 3,987 _____

Rounding to the Nearest Thousand (4 Digit)

Round each number to the nearest thousand

Worksheet # 18

1) 5,545 _____

2) 1,460 _____

3) 5,103 _____

4) 3,697 _____

5) 4,244 _____

6) 9,576 _____

7) 6,174 _____

8) 7,228 _____

9) 1,136 _____

10) 6,127 _____

11) 9,250 _____

12) 6,397 _____

13) 3,589 _____

14) 6,506 _____

15) 3,457 _____

16) 5,157 _____

17) 1,248 _____

18) 6,110 _____

19) 8,589 _____

20) 9,153 _____

21) 2,982 _____

22) 7,759 _____

23) 9,522 _____

24) 5,668 _____

Rounding to the Nearest Thousand (4 Digit)

Round each number to the nearest thousand

Worksheet # 19

1) 5,033 _____

2) 4,862 _____

3) 8,994 _____

4) 6,703 _____

5) 9,776 _____

6) 7,842 _____

7) 9,677 _____

8) 5,411 _____

9) 9,007 _____

10) 7,900 _____

11) 8,473 _____

12) 7,684 _____

13) 1,913 _____

14) 1,176 _____

15) 6,913 _____

16) 4,710 _____

17) 1,323 _____

18) 6,132 _____

19) 3,217 _____

20) 7,834 _____

21) 6,181 _____

22) 7,237 _____

23) 1,533 _____

24) 9,712 _____

Rounding to the Nearest Thousand (4 Digit)

Round each number to the nearest thousand

Worksheet # 20

1) 7,668 _____

2) 6,360 _____

3) 4,379 _____

4) 7,091 _____

5) 6,744 _____

6) 1,250 _____

7) 1,173 _____

8) 2,652 _____

9) 1,183 _____

10) 5,274 _____

11) 1,048 _____

12) 7,908 _____

13) 7,212 _____

14) 7,900 _____

15) 5,608 _____

16) 6,501 _____

17) 1,735 _____

18) 6,703 _____

19) 7,914 _____

20) 9,897 _____

21) 9,671 _____

22) 2,329 _____

23) 3,182 _____

24) 4,291 _____

22

Rounding to the Nearest Thousand (4 Digit)

Round each number to the nearest thousand

Worksheet # 21

1) 6,737 _____

2) 9,893 _____

3) 2,383 _____

4) 3,777 _____

5) 5,125 _____

6) 1,667 _____

7) 2,902 _____

8) 6,359 _____

9) 6,337 _____

10) 9,573 _____

11) 7,331 _____

12) 3,497 _____

13) 4,812 _____

14) 3,863 _____

15) 4,080 _____

16) 7,610 _____

17) 6,160 _____

18) 2,685 _____

19) 5,249 _____

20) 5,621 _____

21) 4,108 _____

22) 1,858 _____

23) 4,350 _____

24) 1,806 _____

Rounding to the Nearest Thousand (4 Digit)

Round each number to the nearest thousand

Worksheet # 22

1) 8,591 _____

2) 5,838 _____

3) 2,967 _____

4) 9,703 _____

5) 6,011 _____

6) 6,861 _____

7) 9,224 _____

8) 9,857 _____

9) 1,586 _____

10) 5,638 _____

11) 3,874 _____

12) 5,926 _____

13) 6,804 _____

14) 4,564 _____

15) 4,281 _____

16) 2,257 _____

17) 4,379 _____

18) 7,674 _____

19) 1,552 _____

20) 9,227 _____

21) 2,480 _____

22) 8,841 _____

23) 3,265 _____

24) 3,729 _____

Rounding to the Nearest Thousand (4 Digit)

Round each number to the nearest thousand

Worksheet # 23

1) 5,683 _____

2) 9,411 _____

3) 3,883 _____

4) 1,300 _____

5) 6,459 _____

6) 1,268 _____

7) 6,267 _____

8) 2,953 _____

9) 6,677 _____

10) 1,693 _____

11) 1,052 _____

12) 1,062 _____

13) 8,640 _____

14) 9,028 _____

15) 9,823 _____

16) 2,365 _____

17) 5,728 _____

18) 1,223 _____

19) 2,475 _____

20) 7,050 _____

21) 3,200 _____

22) 8,518 _____

23) 5,631 _____

24) 3,912 _____

Rounding to the Nearest Thousand (4 Digit)

Round each number to the nearest thousand

Worksheet # 24

1) 3,039 _____

2) 5,382 _____

3) 3,755 _____

4) 1,490 _____

5) 7,159 _____

6) 5,119 _____

7) 6,151 _____

8) 9,714 _____

9) 5,266 _____

10) 9,608 _____

11) 6,322 _____

12) 2,671 _____

13) 2,521 _____

14) 1,056 _____

15) 1,723 _____

16) 3,930 _____

17) 5,350 _____

18) 8,073 _____

19) 5,830 _____

20) 9,278 _____

21) 4,829 _____

22) 5,521 _____

23) 3,272 _____

24) 7,983 _____

Rounding to the Nearest Thousand (4 Digit)

Round each number to the nearest thousand

Worksheet # 25

1) 8,725 _____

2) 1,442 _____

3) 5,244 _____

4) 5,071 _____

5) 4,546 _____

6) 1,255 _____

7) 4,926 _____

8) 6,011 _____

9) 8,735 _____

10) 8,423 _____

11) 4,456 _____

12) 6,752 _____

13) 6,703 _____

14) 6,910 _____

15) 4,219 _____

16) 7,903 _____

17) 1,801 _____

18) 9,468 _____

19) 7,163 _____

20) 8,352 _____

21) 9,825 _____

22) 7,254 _____

23) 5,836 _____

24) 3,128 _____

Rounding to the Nearest Thousand (4 Digit)

Round each number to the nearest thousand

Worksheet # 26

1) 9,073 _____

2) 5,064 _____

3) 7,186 _____

4) 8,478 _____

5) 6,058 _____

6) 6,023 _____

7) 9,110 _____

8) 7,797 _____

9) 1,896 _____

10) 8,971 _____

11) 8,112 _____

12) 4,013 _____

13) 9,000 _____

14) 8,869 _____

15) 5,222 _____

16) 6,063 _____

17) 4,155 _____

18) 6,781 _____

19) 6,576 _____

20) 3,508 _____

21) 8,494 _____

22) 2,316 _____

23) 5,993 _____

24) 6,980 _____

28

Rounding to the Nearest Thousand (4 Digit)

Round each number to the nearest thousand

Worksheet # 27

1) 4,671 _____

2) 4,549 _____

3) 6,344 _____

4) 5,175 _____

5) 4,358 _____

6) 3,525 _____

7) 2,934 _____

8) 4,699 _____

9) 6,275 _____

10) 9,374 _____

11) 2,539 _____

12) 8,600 _____

13) 7,098 _____

14) 9,927 _____

15) 1,106 _____

16) 1,220 _____

17) 4,615 _____

18) 1,266 _____

19) 2,572 _____

20) 3,687 _____

21) 3,075 _____

22) 9,978 _____

23) 2,773 _____

24) 4,521 _____

29

Rounding to the Nearest Thousand (4 Digit)

Round each number to the nearest thousand

Worksheet # 28

1) 7,140 _____

2) 6,565 _____

3) 3,039 _____

4) 4,928 _____

5) 8,215 _____

6) 8,681 _____

7) 2,653 _____

8) 5,749 _____

9) 2,393 _____

10) 9,444 _____

11) 1,457 _____

12) 4,544 _____

13) 7,387 _____

14) 6,205 _____

15) 1,566 _____

16) 8,092 _____

17) 5,874 _____

18) 9,508 _____

19) 2,127 _____

20) 4,730 _____

21) 3,955 _____

22) 4,853 _____

23) 1,585 _____

24) 1,647 _____

Rounding to the Nearest Thousand (4 Digit)

Round each number to the nearest thousand

Worksheet # 29

1) 5,675 _____

2) 1,224 _____

3) 3,309 _____

4) 4,833 _____

5) 7,660 _____

6) 8,178 _____

7) 6,502 _____

8) 3,254 _____

9) 9,750 _____

10) 7,416 _____

11) 3,401 _____

12) 6,330 _____

13) 8,868 _____

14) 3,140 _____

15) 7,414 _____

16) 6,374 _____

17) 9,249 _____

18) 6,488 _____

19) 6,066 _____

20) 9,131 _____

21) 4,073 _____

22) 4,256 _____

23) 1,338 _____

24) 7,140 _____

Rounding to the Nearest Thousand (4 Digit)

Round each number to the nearest thousand

Worksheet # 30

1) 1,553 _____

2) 5,313 _____

3) 4,168 _____

4) 9,656 _____

5) 7,379 _____

6) 5,305 _____

7) 7,412 _____

8) 8,851 _____

9) 4,085 _____

10) 8,112 _____

11) 3,195 _____

12) 1,260 _____

13) 5,005 _____

14) 6,641 _____

15) 8,003 _____

16) 6,011 _____

17) 8,893 _____

18) 3,466 _____

19) 9,040 _____

20) 9,758 _____

21) 1,729 _____

22) 1,501 _____

23) 7,728 _____

24) 5,704 _____

Rounding to the Nearest Thousand (4 Digit)

Round each number to the nearest thousand

Worksheet # 31

1) 9,025 _____

2) 2,781 _____

3) 6,767 _____

4) 4,291 _____

5) 6,489 _____

6) 7,081 _____

7) 4,445 _____

8) 9,259 _____

9) 9,684 _____

10) 1,142 _____

11) 5,574 _____

12) 8,319 _____

13) 5,253 _____

14) 5,246 _____

15) 4,600 _____

16) 7,550 _____

17) 5,991 _____

18) 5,738 _____

19) 5,821 _____

20) 4,462 _____

21) 9,039 _____

22) 5,898 _____

23) 6,388 _____

24) 1,964 _____

Rounding to the Nearest Thousand (4 Digit)

Round each number to the nearest thousand

Worksheet # 32

1) 6,209 _____

2) 3,729 _____

3) 8,918 _____

4) 1,845 _____

5) 8,401 _____

6) 3,923 _____

7) 5,149 _____

8) 5,628 _____

9) 2,917 _____

10) 5,864 _____

11) 2,035 _____

12) 2,899 _____

13) 8,701 _____

14) 2,355 _____

15) 1,815 _____

16) 6,057 _____

17) 4,095 _____

18) 5,243 _____

19) 7,613 _____

20) 8,256 _____

21) 4,178 _____

22) 2,970 _____

23) 7,290 _____

24) 5,100 _____

Rounding to the Nearest Thousand (4 Digit)

Round each number to the nearest thousand

Worksheet # 33

1) 6,396 _____

2) 5,835 _____

3) 3,533 _____

4) 4,661 _____

5) 7,363 _____

6) 7,311 _____

7) 6,289 _____

8) 8,551 _____

9) 4,200 _____

10) 1,084 _____

11) 1,075 _____

12) 2,710 _____

13) 4,292 _____

14) 5,437 _____

15) 2,538 _____

16) 9,623 _____

17) 4,023 _____

18) 7,639 _____

19) 3,713 _____

20) 7,887 _____

21) 4,472 _____

22) 8,943 _____

23) 1,277 _____

24) 8,777 _____

Rounding to the Nearest Thousand (4 Digit)

Round each number to the nearest thousand

Worksheet # 34

1) 7,833 _____

2) 5,645 _____

3) 9,154 _____

4) 1,879 _____

5) 7,081 _____

6) 5,135 _____

7) 1,829 _____

8) 1,365 _____

9) 8,843 _____

10) 8,174 _____

11) 3,538 _____

12) 2,977 _____

13) 1,630 _____

14) 4,162 _____

15) 1,564 _____

16) 5,761 _____

17) 8,027 _____

18) 1,679 _____

19) 8,446 _____

20) 2,132 _____

21) 9,242 _____

22) 2,680 _____

23) 9,771 _____

24) 3,709 _____

Rounding to the Nearest Thousand (4 Digit)

Round each number to the nearest thousand

Worksheet # 35

1)   1,803   _____

2)   7,615   _____

3)   4,929   _____

4)   6,084   _____

5)   2,183   _____

6)   7,208   _____

7)   3,184   _____

8)   8,286   _____

9)   3,257   _____

10)   1,788   _____

11)   7,479   _____

12)   4,164   _____

13)   4,628   _____

14)   5,200   _____

15)   4,320   _____

16)   2,882   _____

17)   8,409   _____

18)   9,597   _____

19)   2,488   _____

20)   6,127   _____

21)   3,935   _____

22)   4,456   _____

23)   8,024   _____

24)   2,069   _____

Rounding to the Nearest Thousand (4 Digit)

Round each number to the nearest thousand

Worksheet # 36

1) 5,260 _____

2) 6,765 _____

3) 6,392 _____

4) 9,506 _____

5) 9,981 _____

6) 1,946 _____

7) 1,912 _____

8) 7,310 _____

9) 7,211 _____

10) 8,903 _____

11) 1,999 _____

12) 8,398 _____

13) 4,918 _____

14) 4,596 _____

15) 5,616 _____

16) 3,346 _____

17) 3,302 _____

18) 2,787 _____

19) 2,105 _____

20) 8,833 _____

21) 1,556 _____

22) 9,438 _____

23) 4,240 _____

24) 4,350 _____

Rounding to the Nearest Thousand (4 Digit)

Round each number to the nearest thousand

Worksheet # 37

1) 2,627 _____

2) 9,409 _____

3) 9,095 _____

4) 3,894 _____

5) 2,509 _____

6) 6,891 _____

7) 1,955 _____

8) 6,592 _____

9) 3,559 _____

10) 1,574 _____

11) 1,412 _____

12) 8,273 _____

13) 3,628 _____

14) 4,725 _____

15) 3,783 _____

16) 8,721 _____

17) 5,139 _____

18) 5,598 _____

19) 6,952 _____

20) 6,780 _____

21) 1,386 _____

22) 7,112 _____

23) 8,095 _____

24) 2,083 _____

Rounding to the Nearest Thousand (4 Digit)

Round each number to the nearest thousand

Worksheet # 38

1) 7,775 _____

2) 9,184 _____

3) 7,463 _____

4) 8,197 _____

5) 6,229 _____

6) 4,460 _____

7) 2,564 _____

8) 8,524 _____

9) 3,814 _____

10) 7,316 _____

11) 4,747 _____

12) 8,557 _____

13) 8,225 _____

14) 5,791 _____

15) 6,483 _____

16) 4,236 _____

17) 4,312 _____

18) 2,262 _____

19) 8,599 _____

20) 9,126 _____

21) 3,088 _____

22) 7,198 _____

23) 8,819 _____

24) 8,746 _____

40

Rounding to the Nearest Thousand (4 Digit)

Round each number to the nearest thousand

Worksheet # 39

1) 7,579 _____

2) 4,633 _____

3) 6,572 _____

4) 7,035 _____

5) 2,107 _____

6) 1,174 _____

7) 3,259 _____

8) 9,961 _____

9) 8,424 _____

10) 3,544 _____

11) 2,362 _____

12) 8,219 _____

13) 7,701 _____

14) 8,197 _____

15) 4,820 _____

16) 8,126 _____

17) 6,791 _____

18) 5,657 _____

19) 7,608 _____

20) 7,215 _____

21) 9,954 _____

22) 9,503 _____

23) 8,308 _____

24) 6,967 _____

Rounding to the Nearest Thousand (4 Digit)

Round each number to the nearest thousand

Worksheet # 40

1) 2,322 _____

2) 4,816 _____

3) 5,915 _____

4) 7,545 _____

5) 3,728 _____

6) 3,408 _____

7) 6,222 _____

8) 3,272 _____

9) 5,527 _____

10) 5,378 _____

11) 8,435 _____

12) 7,421 _____

13) 6,560 _____

14) 9,189 _____

15) 6,088 _____

16) 5,718 _____

17) 2,380 _____

18) 6,343 _____

19) 9,250 _____

20) 1,473 _____

21) 8,934 _____

22) 4,870 _____

23) 7,211 _____

24) 3,441 _____

42

Rounding to the Nearest Thousand (4 Digit)

Round each number to the nearest thousand

Worksheet # 41

1) 2,275 _____

2) 8,924 _____

3) 2,250 _____

4) 3,620 _____

5) 8,281 _____

6) 4,644 _____

7) 8,697 _____

8) 8,840 _____

9) 9,916 _____

10) 2,922 _____

11) 8,012 _____

12) 4,646 _____

13) 5,986 _____

14) 8,439 _____

15) 2,340 _____

16) 5,351 _____

17) 2,440 _____

18) 4,827 _____

19) 2,268 _____

20) 5,536 _____

21) 6,458 _____

22) 4,824 _____

23) 1,077 _____

24) 3,772 _____

Rounding to the Nearest Thousand (4 Digit)

Round each number to the nearest thousand

Worksheet # 42

1) 1,959 _____

2) 9,877 _____

3) 2,185 _____

4) 7,472 _____

5) 9,963 _____

6) 3,236 _____

7) 5,054 _____

8) 2,934 _____

9) 6,241 _____

10) 7,879 _____

11) 4,088 _____

12) 1,829 _____

13) 3,166 _____

14) 9,649 _____

15) 5,521 _____

16) 7,402 _____

17) 8,265 _____

18) 1,166 _____

19) 7,191 _____

20) 5,652 _____

21) 4,133 _____

22) 6,666 _____

23) 4,094 _____

24) 9,797 _____

Rounding to the Nearest Thousand (4 Digit)

Round each number to the nearest thousand

Worksheet # 43

1) 5,146 _____

2) 3,073 _____

3) 7,410 _____

4) 8,523 _____

5) 4,551 _____

6) 3,705 _____

7) 1,010 _____

8) 7,773 _____

9) 9,353 _____

10) 4,037 _____

11) 5,034 _____

12) 9,610 _____

13) 8,121 _____

14) 6,707 _____

15) 1,254 _____

16) 9,787 _____

17) 2,127 _____

18) 4,638 _____

19) 3,318 _____

20) 1,226 _____

21) 9,182 _____

22) 6,372 _____

23) 9,059 _____

24) 5,177 _____

Rounding to the Nearest Thousand (4 Digit)

Round each number to the nearest thousand

Worksheet # 44

1) 2,720 _____

2) 9,042 _____

3) 1,712 _____

4) 4,245 _____

5) 1,202 _____

6) 3,951 _____

7) 4,761 _____

8) 8,021 _____

9) 8,328 _____

10) 9,782 _____

11) 4,801 _____

12) 3,778 _____

13) 1,931 _____

14) 1,975 _____

15) 5,416 _____

16) 9,210 _____

17) 9,810 _____

18) 5,610 _____

19) 4,641 _____

20) 3,765 _____

21) 2,379 _____

22) 5,015 _____

23) 5,523 _____

24) 9,307 _____

46

Rounding to the Nearest Thousand (4 Digit)

Round each number to the nearest thousand

Worksheet # 45

1) 6,398 _____

2) 8,654 _____

3) 9,664 _____

4) 6,201 _____

5) 5,405 _____

6) 4,205 _____

7) 4,193 _____

8) 8,929 _____

9) 7,206 _____

10) 3,248 _____

11) 3,962 _____

12) 3,120 _____

13) 3,674 _____

14) 3,638 _____

15) 4,703 _____

16) 8,445 _____

17) 4,148 _____

18) 3,675 _____

19) 1,075 _____

20) 1,000 _____

21) 2,076 _____

22) 5,297 _____

23) 9,503 _____

24) 1,004 _____

47

Rounding to the Nearest Thousand (4 Digit)

Round each number to the nearest thousand

Worksheet # 46

1) 9,120 _____

2) 1,520 _____

3) 7,891 _____

4) 5,354 _____

5) 3,636 _____

6) 2,456 _____

7) 4,767 _____

8) 6,028 _____

9) 3,222 _____

10) 8,504 _____

11) 3,903 _____

12) 7,278 _____

13) 9,311 _____

14) 4,453 _____

15) 1,881 _____

16) 5,984 _____

17) 5,112 _____

18) 4,253 _____

19) 5,837 _____

20) 1,812 _____

21) 4,815 _____

22) 5,505 _____

23) 6,151 _____

24) 9,620 _____

Rounding to the Nearest Thousand (4 Digit)

Round each number to the nearest thousand

Worksheet # 47

1) 3,893 _____

2) 1,300 _____

3) 3,893 _____

4) 7,874 _____

5) 7,605 _____

6) 2,171 _____

7) 1,480 _____

8) 2,550 _____

9) 6,051 _____

10) 1,638 _____

11) 7,086 _____

12) 9,311 _____

13) 4,111 _____

14) 1,782 _____

15) 6,802 _____

16) 9,240 _____

17) 9,674 _____

18) 7,157 _____

19) 3,907 _____

20) 2,088 _____

21) 5,925 _____

22) 5,031 _____

23) 1,282 _____

24) 3,807 _____

Rounding to the Nearest Thousand (4 Digit)

Round each number to the nearest thousand

Worksheet # 48

1) 8,600 _____

2) 4,012 _____

3) 5,657 _____

4) 2,716 _____

5) 7,830 _____

6) 6,698 _____

7) 4,687 _____

8) 8,096 _____

9) 7,988 _____

10) 3,061 _____

11) 9,296 _____

12) 1,873 _____

13) 1,917 _____

14) 7,409 _____

15) 4,307 _____

16) 5,720 _____

17) 9,995 _____

18) 5,918 _____

19) 5,867 _____

20) 3,651 _____

21) 6,573 _____

22) 4,049 _____

23) 5,326 _____

24) 3,732 _____

Rounding to the Nearest Thousand (4 Digit)

Round each number to the nearest thousand

Worksheet # 49

1) 1,217 _____

2) 9,223 _____

3) 7,764 _____

4) 6,086 _____

5) 3,996 _____

6) 6,559 _____

7) 2,940 _____

8) 2,254 _____

9) 6,363 _____

10) 1,696 _____

11) 4,454 _____

12) 9,837 _____

13) 5,229 _____

14) 8,564 _____

15) 8,501 _____

16) 5,421 _____

17) 2,241 _____

18) 4,204 _____

19) 7,954 _____

20) 9,052 _____

21) 6,714 _____

22) 9,053 _____

23) 1,046 _____

24) 5,323 _____

Rounding to the Nearest Thousand (4 Digit)

Round each number to the nearest thousand

Worksheet # 50

1) 2,471 _____

2) 8,096 _____

3) 4,396 _____

4) 5,076 _____

5) 4,331 _____

6) 6,272 _____

7) 2,445 _____

8) 9,434 _____

9) 7,324 _____

10) 2,251 _____

11) 3,502 _____

12) 8,149 _____

13) 8,794 _____

14) 1,722 _____

15) 5,351 _____

16) 3,951 _____

17) 3,228 _____

18) 2,731 _____

19) 4,008 _____

20) 6,314 _____

21) 8,444 _____

22) 9,646 _____

23) 9,082 _____

24) 8,414 _____

Answer Key

Answer Key

Worksheet # 1

1) 4,000   2) 3,000   3) 3,000   4) 5,000   5) 6,000   6) 4,000   7) 9,000   8) 6,000

9) 7,000   10) 1,000   11) 6,000   12) 7,000   13) 8,000   14) 3,000   15) 8,000   16) 8,000

17) 8,000   18) 3,000   19) 2,000   20) 3,000   21) 5,000   22) 7,000   23) 4,000   24) 3,000

Worksheet # 2

1) 7,000   2) 9,000   3) 5,000   4) 6,000   5) 9,000   6) 7,000   7) 8,000   8) 6,000

9) 7,000   10) 10,000   11) 6,000   12) 5,000   13) 4,000   14) 8,000   15) 8,000   16) 5,000

17) 2,000   18) 2,000   19) 3,000   20) 9,000   21) 2,000   22) 10,000   23) 6,000   24) 3,000

Worksheet # 3

1) 6,000   2) 2,000   3) 4,000   4) 2,000   5) 2,000   6) 9,000   7) 8,000   8) 4,000

9) 3,000   10) 2,000   11) 6,000   12) 1,000   13) 5,000   14) 5,000   15) 1,000   16) 2,000

17) 2,000   18) 6,000   19) 8,000   20) 8,000   21) 5,000   22) 5,000   23) 2,000   24) 10,000

Worksheet # 4

1) 4,000   2) 6,000   3) 2,000   4) 2,000   5) 1,000   6) 5,000   7) 9,000   8) 9,000

9) 2,000   10) 1,000   11) 4,000   12) 6,000   13) 5,000   14) 8,000   15) 7,000   16) 3,000

17) 9,000   18) 9,000   19) 4,000   20) 7,000   21) 5,000   22) 4,000   23) 9,000   24) 4,000

Worksheet # 5

1) 6,000   2) 9,000   3) 6,000   4) 6,000   5) 4,000   6) 9,000   7) 9,000   8) 3,000

9) 7,000   10) 6,000   11) 8,000   12) 2,000   13) 7,000   14) 6,000   15) 2,000   16) 4,000

17) 7,000   18) 7,000   19) 3,000   20) 8,000   21) 4,000   22) 10,000   23) 2,000   24) 7,000

Worksheet # 6

1) 4,000   2) 5,000   3) 2,000   4) 2,000   5) 2,000   6) 5,000   7) 4,000   8) 9,000

9) 2,000   10) 5,000   11) 8,000   12) 3,000   13) 8,000   14) 1,000   15) 5,000   16) 7,000

17) 8,000   18) 6,000   19) 3,000   20) 10,000   21) 8,000   22) 4,000   23) 9,000   24) 6,000

Answer Key

Worksheet # 7

1) 9,000   2) 5,000   3) 10,000   4) 8,000   5) 8,000   6) 10,000   7) 8,000   8) 10,000

9) 10,000   10) 2,000   11) 9,000   12) 9,000   13) 1,000   14) 7,000   15) 2,000   16) 7,000

17) 2,000   18) 2,000   19) 9,000   20) 7,000   21) 1,000   22) 2,000   23) 2,000   24) 3,000

Worksheet # 8

1) 8,000   2) 7,000   3) 8,000   4) 4,000   5) 5,000   6) 2,000   7) 6,000   8) 2,000

9) 8,000   10) 5,000   11) 3,000   12) 2,000   13) 9,000   14) 4,000   15) 10,000   16) 6,000

17) 6,000   18) 2,000   19) 4,000   20) 3,000   21) 4,000   22) 10,000   23) 7,000   24) 3,000

Worksheet # 9

1) 10,000   2) 7,000   3) 5,000   4) 2,000   5) 1,000   6) 10,000   7) 4,000   8) 2,000

9) 5,000   10) 3,000   11) 5,000   12) 3,000   13) 9,000   14) 7,000   15) 3,000   16) 6,000

17) 1,000   18) 10,000   19) 9,000   20) 2,000   21) 10,000   22) 5,000   23) 6,000   24) 1,000

Worksheet # 10

1) 4,000   2) 1,000   3) 7,000   4) 9,000   5) 2,000   6) 4,000   7) 7,000   8) 4,000

9) 8,000   10) 3,000   11) 8,000   12) 10,000   13) 6,000   14) 1,000   15) 1,000   16) 2,000

17) 9,000   18) 3,000   19) 4,000   20) 7,000   21) 9,000   22) 5,000   23) 7,000   24) 7,000

Worksheet # 11

1) 4,000   2) 1,000   3) 8,000   4) 6,000   5) 1,000   6) 7,000   7) 2,000   8) 8,000

9) 6,000   10) 2,000   11) 8,000   12) 3,000   13) 5,000   14) 6,000   15) 8,000   16) 4,000

17) 2,000   18) 10,000   19) 3,000   20) 1,000   21) 3,000   22) 5,000   23) 1,000   24) 3,000

Worksheet # 12

1) 6,000   2) 10,000   3) 4,000   4) 8,000   5) 2,000   6) 4,000   7) 3,000   8) 5,000

9) 5,000   10) 5,000   11) 10,000   12) 10,000   13) 2,000   14) 5,000   15) 1,000   16) 2,000

17) 4,000   18) 1,000   19) 2,000   20) 8,000   21) 5,000   22) 4,000   23) 9,000   24) 2,000

Answer Key

Worksheet # 13

1) 4,000   2) 5,000   3) 3,000   4) 1,000   5) 7,000   6) 3,000   7) 7,000   8) 5,000

9) 6,000   10) 6,000   11) 6,000   12) 9,000   13) 3,000   14) 6,000   15) 9,000   16) 3,000

17) 5,000   18) 6,000   19) 6,000   20) 8,000   21) 9,000   22) 5,000   23) 5,000   24) 6,000

Worksheet # 14

1) 7,000   2) 2,000   3) 3,000   4) 6,000   5) 4,000   6) 7,000   7) 10,000   8) 3,000

9) 1,000   10) 7,000   11) 9,000   12) 7,000   13) 9,000   14) 4,000   15) 9,000   16) 8,000

17) 4,000   18) 5,000   19) 10,000   20) 3,000   21) 3,000   22) 9,000   23) 1,000   24) 9,000

Worksheet # 15

1) 2,000   2) 5,000   3) 5,000   4) 4,000   5) 5,000   6) 6,000   7) 2,000   8) 4,000

9) 6,000   10) 6,000   11) 3,000   12) 2,000   13) 10,000   14) 5,000   15) 3,000   16) 10,000

17) 5,000   18) 4,000   19) 3,000   20) 4,000   21) 2,000   22) 9,000   23) 4,000   24) 10,000

Worksheet # 16

1) 4,000   2) 8,000   3) 7,000   4) 3,000   5) 2,000   6) 9,000   7) 9,000   8) 10,000

9) 10,000   10) 3,000   11) 4,000   12) 2,000   13) 6,000   14) 2,000   15) 2,000   16) 7,000

17) 7,000   18) 2,000   19) 9,000   20) 8,000   21) 2,000   22) 7,000   23) 10,000   24) 7,000

Worksheet # 17

1) 1,000   2) 5,000   3) 7,000   4) 4,000   5) 6,000   6) 6,000   7) 5,000   8) 4,000

9) 6,000   10) 2,000   11) 1,000   12) 5,000   13) 3,000   14) 7,000   15) 4,000   16) 1,000

17) 8,000   18) 5,000   19) 8,000   20) 10,000   21) 4,000   22) 6,000   23) 2,000   24) 4,000

Worksheet # 18

1) 6,000   2) 1,000   3) 5,000   4) 4,000   5) 4,000   6) 10,000   7) 6,000   8) 7,000

9) 1,000   10) 6,000   11) 9,000   12) 6,000   13) 4,000   14) 7,000   15) 3,000   16) 5,000

17) 1,000   18) 6,000   19) 9,000   20) 9,000   21) 3,000   22) 8,000   23) 10,000   24) 6,000

Answer Key

Worksheet # 19

1) 5,000   2) 5,000   3) 9,000   4) 7,000   5) 10,000   6) 8,000   7) 10,000   8) 5,000

9) 9,000   10) 8,000   11) 8,000   12) 8,000   13) 2,000   14) 1,000   15) 7,000   16) 5,000

17) 1,000   18) 6,000   19) 3,000   20) 8,000   21) 6,000   22) 7,000   23) 2,000   24) 10,000

Worksheet # 20

1) 8,000   2) 6,000   3) 4,000   4) 7,000   5) 7,000   6) 1,000   7) 1,000   8) 3,000

9) 1,000   10) 5,000   11) 1,000   12) 8,000   13) 7,000   14) 8,000   15) 6,000   16) 7,000

17) 2,000   18) 7,000   19) 8,000   20) 10,000   21) 10,000   22) 2,000   23) 3,000   24) 4,000

Worksheet # 21

1) 7,000   2) 10,000   3) 2,000   4) 4,000   5) 5,000   6) 2,000   7) 3,000   8) 6,000

9) 6,000   10) 10,000   11) 7,000   12) 3,000   13) 5,000   14) 4,000   15) 4,000   16) 8,000

17) 6,000   18) 3,000   19) 5,000   20) 6,000   21) 4,000   22) 2,000   23) 4,000   24) 2,000

Worksheet # 22

1) 9,000   2) 6,000   3) 3,000   4) 10,000   5) 6,000   6) 7,000   7) 9,000   8) 10,000

9) 2,000   10) 6,000   11) 4,000   12) 6,000   13) 7,000   14) 5,000   15) 4,000   16) 2,000

17) 4,000   18) 8,000   19) 2,000   20) 9,000   21) 2,000   22) 9,000   23) 3,000   24) 4,000

Worksheet # 23

1) 6,000   2) 9,000   3) 4,000   4) 1,000   5) 6,000   6) 1,000   7) 6,000   8) 3,000

9) 7,000   10) 2,000   11) 1,000   12) 1,000   13) 9,000   14) 9,000   15) 10,000   16) 2,000

17) 6,000   18) 1,000   19) 2,000   20) 7,000   21) 3,000   22) 9,000   23) 6,000   24) 4,000

Worksheet # 24

1) 3,000   2) 5,000   3) 4,000   4) 1,000   5) 7,000   6) 5,000   7) 6,000   8) 10,000

9) 5,000   10) 10,000   11) 6,000   12) 3,000   13) 3,000   14) 1,000   15) 2,000   16) 4,000

17) 5,000   18) 8,000   19) 6,000   20) 9,000   21) 5,000   22) 6,000   23) 3,000   24) 8,000

Answer Key

Worksheet # 25

1) 9,000	2) 1,000	3) 5,000	4) 5,000	5) 5,000	6) 1,000	7) 5,000	8) 6,000
9) 9,000	10) 8,000	11) 4,000	12) 7,000	13) 7,000	14) 7,000	15) 4,000	16) 8,000
17) 2,000	18) 9,000	19) 7,000	20) 8,000	21) 10,000	22) 7,000	23) 6,000	24) 3,000

Worksheet # 26

1) 9,000	2) 5,000	3) 7,000	4) 8,000	5) 6,000	6) 6,000	7) 9,000	8) 8,000
9) 2,000	10) 9,000	11) 8,000	12) 4,000	13) 9,000	14) 9,000	15) 5,000	16) 6,000
17) 4,000	18) 7,000	19) 7,000	20) 4,000	21) 8,000	22) 2,000	23) 6,000	24) 7,000

Worksheet # 27

1) 5,000	2) 5,000	3) 6,000	4) 5,000	5) 4,000	6) 4,000	7) 3,000	8) 5,000
9) 6,000	10) 9,000	11) 3,000	12) 9,000	13) 7,000	14) 10,000	15) 1,000	16) 1,000
17) 5,000	18) 1,000	19) 3,000	20) 4,000	21) 3,000	22) 10,000	23) 3,000	24) 5,000

Worksheet # 28

1) 7,000	2) 7,000	3) 3,000	4) 5,000	5) 8,000	6) 9,000	7) 3,000	8) 6,000
9) 2,000	10) 9,000	11) 1,000	12) 5,000	13) 7,000	14) 6,000	15) 2,000	16) 8,000
17) 6,000	18) 10,000	19) 2,000	20) 5,000	21) 4,000	22) 5,000	23) 2,000	24) 2,000

Worksheet # 29

1) 6,000	2) 1,000	3) 3,000	4) 5,000	5) 8,000	6) 8,000	7) 7,000	8) 3,000
9) 10,000	10) 7,000	11) 3,000	12) 6,000	13) 9,000	14) 3,000	15) 7,000	16) 6,000
17) 9,000	18) 6,000	19) 6,000	20) 9,000	21) 4,000	22) 4,000	23) 1,000	24) 7,000

Worksheet # 30

1) 2,000	2) 5,000	3) 4,000	4) 10,000	5) 7,000	6) 5,000	7) 7,000	8) 9,000
9) 4,000	10) 8,000	11) 3,000	12) 1,000	13) 5,000	14) 7,000	15) 8,000	16) 6,000
17) 9,000	18) 3,000	19) 9,000	20) 10,000	21) 2,000	22) 2,000	23) 8,000	24) 6,000

Answer Key

Worksheet # 31

1) 9,000	2) 3,000	3) 7,000	4) 4,000	5) 6,000	6) 7,000	7) 4,000	8) 9,000
9) 10,000	10) 1,000	11) 6,000	12) 8,000	13) 5,000	14) 5,000	15) 5,000	16) 8,000
17) 6,000	18) 6,000	19) 6,000	20) 4,000	21) 9,000	22) 6,000	23) 6,000	24) 2,000

Worksheet # 32

1) 6,000	2) 4,000	3) 9,000	4) 2,000	5) 8,000	6) 4,000	7) 5,000	8) 6,000
9) 3,000	10) 6,000	11) 2,000	12) 3,000	13) 9,000	14) 2,000	15) 2,000	16) 6,000
17) 4,000	18) 5,000	19) 8,000	20) 8,000	21) 4,000	22) 3,000	23) 7,000	24) 5,000

Worksheet # 33

1) 6,000	2) 6,000	3) 4,000	4) 5,000	5) 7,000	6) 7,000	7) 6,000	8) 9,000
9) 4,000	10) 1,000	11) 1,000	12) 3,000	13) 4,000	14) 5,000	15) 3,000	16) 10,000
17) 4,000	18) 8,000	19) 4,000	20) 8,000	21) 4,000	22) 9,000	23) 1,000	24) 9,000

Worksheet # 34

1) 8,000	2) 6,000	3) 9,000	4) 2,000	5) 7,000	6) 5,000	7) 2,000	8) 1,000
9) 9,000	10) 8,000	11) 4,000	12) 3,000	13) 2,000	14) 4,000	15) 2,000	16) 6,000
17) 8,000	18) 2,000	19) 8,000	20) 2,000	21) 9,000	22) 3,000	23) 10,000	24) 4,000

Worksheet # 35

1) 2,000	2) 8,000	3) 5,000	4) 6,000	5) 2,000	6) 7,000	7) 3,000	8) 8,000
9) 3,000	10) 2,000	11) 7,000	12) 4,000	13) 5,000	14) 5,000	15) 4,000	16) 3,000
17) 8,000	18) 10,000	19) 2,000	20) 6,000	21) 4,000	22) 4,000	23) 8,000	24) 2,000

Worksheet # 36

1) 5,000	2) 7,000	3) 6,000	4) 10,000	5) 10,000	6) 2,000	7) 2,000	8) 7,000
9) 7,000	10) 9,000	11) 2,000	12) 8,000	13) 5,000	14) 5,000	15) 6,000	16) 3,000
17) 3,000	18) 3,000	19) 2,000	20) 9,000	21) 2,000	22) 9,000	23) 4,000	24) 4,000

Answer Key

Worksheet # 37
1) 3,000   2) 9,000   3) 9,000   4) 4,000   5) 3,000   6) 7,000   7) 2,000   8) 7,000
9) 4,000   10) 2,000   11) 1,000   12) 8,000   13) 4,000   14) 5,000   15) 4,000   16) 9,000
17) 5,000   18) 6,000   19) 7,000   20) 7,000   21) 1,000   22) 7,000   23) 8,000   24) 2,000

Worksheet # 38
1) 8,000   2) 9,000   3) 7,000   4) 8,000   5) 6,000   6) 4,000   7) 3,000   8) 9,000
9) 4,000   10) 7,000   11) 5,000   12) 9,000   13) 8,000   14) 6,000   15) 6,000   16) 4,000
17) 4,000   18) 2,000   19) 9,000   20) 9,000   21) 3,000   22) 7,000   23) 9,000   24) 9,000

Worksheet # 39
1) 8,000   2) 5,000   3) 7,000   4) 7,000   5) 2,000   6) 1,000   7) 3,000   8) 10,000
9) 8,000   10) 4,000   11) 2,000   12) 8,000   13) 8,000   14) 8,000   15) 5,000   16) 8,000
17) 7,000   18) 6,000   19) 8,000   20) 7,000   21) 10,000   22) 10,000   23) 8,000   24) 7,000

Worksheet # 40
1) 2,000   2) 5,000   3) 6,000   4) 8,000   5) 4,000   6) 3,000   7) 6,000   8) 3,000
9) 6,000   10) 5,000   11) 8,000   12) 7,000   13) 7,000   14) 9,000   15) 6,000   16) 6,000
17) 2,000   18) 6,000   19) 9,000   20) 1,000   21) 9,000   22) 5,000   23) 7,000   24) 3,000

Worksheet # 41
1) 2,000   2) 9,000   3) 2,000   4) 4,000   5) 8,000   6) 5,000   7) 9,000   8) 9,000
9) 10,000   10) 3,000   11) 8,000   12) 5,000   13) 6,000   14) 8,000   15) 2,000   16) 5,000
17) 2,000   18) 5,000   19) 2,000   20) 6,000   21) 6,000   22) 5,000   23) 1,000   24) 4,000

Worksheet # 42
1) 2,000   2) 10,000   3) 2,000   4) 7,000   5) 10,000   6) 3,000   7) 5,000   8) 3,000
9) 6,000   10) 8,000   11) 4,000   12) 2,000   13) 3,000   14) 10,000   15) 6,000   16) 7,000
17) 8,000   18) 1,000   19) 7,000   20) 6,000   21) 4,000   22) 7,000   23) 4,000   24) 10,000

Answer Key

Worksheet # 43

1) 5,000   2) 3,000   3) 7,000   4) 9,000   5) 5,000   6) 4,000   7) 1,000   8) 8,000

9) 9,000   10) 4,000   11) 5,000   12) 10,000   13) 8,000   14) 7,000   15) 1,000   16) 10,000

17) 2,000   18) 5,000   19) 3,000   20) 1,000   21) 9,000   22) 6,000   23) 9,000   24) 5,000

Worksheet # 44

1) 3,000   2) 9,000   3) 2,000   4) 4,000   5) 1,000   6) 4,000   7) 5,000   8) 8,000

9) 8,000   10) 10,000   11) 5,000   12) 4,000   13) 2,000   14) 2,000   15) 5,000   16) 9,000

17) 10,000   18) 6,000   19) 5,000   20) 4,000   21) 2,000   22) 5,000   23) 6,000   24) 9,000

Worksheet # 45

1) 6,000   2) 9,000   3) 10,000   4) 6,000   5) 5,000   6) 4,000   7) 4,000   8) 9,000

9) 7,000   10) 3,000   11) 4,000   12) 3,000   13) 4,000   14) 4,000   15) 5,000   16) 8,000

17) 4,000   18) 4,000   19) 1,000   20) 1,000   21) 2,000   22) 5,000   23) 10,000   24) 1,000

Worksheet # 46

1) 9,000   2) 2,000   3) 8,000   4) 5,000   5) 4,000   6) 2,000   7) 5,000   8) 6,000

9) 3,000   10) 9,000   11) 4,000   12) 7,000   13) 9,000   14) 4,000   15) 2,000   16) 6,000

17) 5,000   18) 4,000   19) 6,000   20) 2,000   21) 5,000   22) 6,000   23) 6,000   24) 10,000

Worksheet # 47

1) 4,000   2) 1,000   3) 4,000   4) 8,000   5) 8,000   6) 2,000   7) 1,000   8) 3,000

9) 6,000   10) 2,000   11) 7,000   12) 9,000   13) 4,000   14) 2,000   15) 7,000   16) 9,000

17) 10,000   18) 7,000   19) 4,000   20) 2,000   21) 6,000   22) 5,000   23) 1,000   24) 4,000

Worksheet # 48

1) 9,000   2) 4,000   3) 6,000   4) 3,000   5) 8,000   6) 7,000   7) 5,000   8) 8,000

9) 8,000   10) 3,000   11) 9,000   12) 2,000   13) 2,000   14) 7,000   15) 4,000   16) 6,000

17) 10,000   18) 6,000   19) 6,000   20) 4,000   21) 7,000   22) 4,000   23) 5,000   24) 4,000

Answer Key

Worksheet # 49

1) 1,000   2) 9,000   3) 8,000   4) 6,000   5) 4,000   6) 7,000   7) 3,000   8) 2,000

9) 6,000   10) 2,000   11) 4,000   12) 10,000   13) 5,000   14) 9,000   15) 9,000   16) 5,000

17) 2,000   18) 4,000   19) 8,000   20) 9,000   21) 7,000   22) 9,000   23) 1,000   24) 5,000

Worksheet # 50

1) 2,000   2) 8,000   3) 4,000   4) 5,000   5) 4,000   6) 6,000   7) 2,000   8) 9,000

9) 7,000   10) 2,000   11) 4,000   12) 8,000   13) 9,000   14) 2,000   15) 5,000   16) 4,000

17) 3,000   18) 3,000   19) 4,000   20) 6,000   21) 8,000   22) 10,000   23) 9,000   24) 8,000